Crazy, Strange, and Disruptive Ideas

What are the bounds of thought? And how do we know if an idea is completely insane, or just a paradigm change?

In this book we explore Insanity, Crazy, Strange, and Disruptive ideas, and just some thoughts and stories.

And what are disruptive ideas or ideas which break currently accepted paradigms?

My goal is for you the reader to expand your minds to think "Outside of the Box" and not immediately condemn or make fun of new ideas just because they sound weird or wacky. Who knows? These new ideas might actually change the world for the better.

Crazy, Strange, and Disruptive Ideas

Crazy, Strange, and Disruptive Ideas

Copyright Page

The book is copyrighted for 2019

Crazy, Strange, and Disruptive Ideas

The Crazy and Out of the Box Series Book 1

By Martin K. Ettington

All Rights Reserved USA 2019

ISBN: 9781086177183

Printed in the United States of America

Crazy, Strange, and Disruptive Ideas

Crazy, Strange, and Disruptive Ideas

Other books by Martin K. Ettington

Spiritual and Metaphysics Books:
Prophecy: A History and How to Guide
God Like Powers and Abilities
Enlightenment for Newbies
Removing Illusions to Find True Happiness
Using the Scientific Method to Study the Paranormal
A Compendium of Metaphysics and How to Guides (Six books together in one volume)
Love from the Heart
The Enlightenment Experience
Learn Your Soul's Purpose
Pursuing Enlightenment
A Modern Man's Search for Truth
Use Intuition and Prophecy to Improve Your Life
The Handbook of Spiritual and Energy Healing

Longevity & Immortality:
Physical Immortality: A History and How to Guide
The Commentaries of Living Immortals
Records of Extremely Long Lived Persons
Enlightenment and Immortality
Longevity Improvements from Science
The 10 Principles of Personal Longevity
Telomeres & Longevity
The Diets and Lifestyles of the Worlds Oldest Peoples
The Longevity Six Books Bundle

Science Fiction:
Out of This Universe
Personal Freedom-Parts 1 & 2
The Psychic Soldier Series:
    Book 1-Himalayan Journey
    Book 2-A Soldier is Born
    Book 3-Fighting For Right
    Book 4-Earth Protector
The Immortality Sci Fi Bundle

The God Like Powers Series:
Human Invisibility
Invulnerability and Shielding
Teleportation
Psychokinesis
Our Energy Body, Auras, and Thoughtforms

The God Like Powers Series—
    Volume 1 Compilation
The Yoga Discovery Series:
Yoga-An Ancient Art Form
Hatha Yoga-Helping you Live Better
Raja Yoga-Through the Ages
The Yoga Discovery Package

Business & Coaching Books:
Creating, Publishing, & Marketing Practitioner Ebooks
Building a Successful Longevity Coaching Business
Why Become a Coach?
The Professional Coaching Success Trilogy
2020-Make Money Writing and Selling Books
The 2020 Handbook of High Paying Work Without a College Degree

Science, Technology, and Misc.
Future Predictions By and Engineer & Seer
The Unusual Science & Technology Bundle
The Real Atlantis-In the Eye of the Sahara
Are Cryptozoological Animals Real or Imaginary?
Real Time Travel Stories From a Psychic Engineer
Removing Limits On Our Consciousness-And Thinking Outside the Box
33 Incredible True Survival Stories
How to Survive Anything: From the Wilderness to Man Made Disasters
All About Mars Journeys and Settlement
Mining the Asteroid Belt

Ancient History
The Real Atlantis-In the Eye of the Sahara
Ancient & Prehistoric Civilizations
Ancient & Prehistoric Civilizations-Book Two
The History of Antediluvian Giants
The Antediluvian History of Earth

Crazy, Strange, and Disruptive Ideas

Ancient Underground Cities and Tunnels
Strange Objects Which Should Not Exist
Strange and Ancient Places in the USA
A Theory of Ancient Prehistory And Giant Aliens
Aliens and Space
Aliens and Secret Technology
Aliens Are Already Among Us
Designing and Building Space Colonies

Humanity and the Universe
All About Moon Bases
All About Mars Journeys and Settlement
The Space and Aliens Six Books Bundle
A Theory of Ancient Prehistory and Giant Aliens
The Space Colonies and Space Structures Coloring Book
All About Asteroids

The Longevity Training Series

(A transcription of the online Multimedia Longevity Coaching Training Program)

The Personal Longevity Training Series-Book1-Long Lived Persons
The Personal Longevity Training Series-Book2-Your Soul's Purpose
The Personal Longevity Training Series-Book3-Enable Your Life Urge
The Personal Longevity Training Series-Book4-Your Spiritual Connection
The Personal Longevity Training Series-Book5-Having Love in Your Heart
The Personal Longevity Training Series-Book6-Energy Body Health
The Personal Longevity Training Series-Book7-The Science of Longevity
The Personal Longevity Training Series-Book8-Physical Body Health
The Personal Longevity Training Series-Book9-Avoiding Accidents
The Personal Longevity Training Series-Book10-Implementing These Principles

The Personal Longevity Training Series-Books One Thru Ten

These books are all available in digital and printed formats from my website and on Amazon, Barnes & Noble, Apple ITunes, and many other sites

My Books Website is: http://mkettingtonbooks.com

Crazy, Strange, and Disruptive Ideas

Signup for our Mailing List to get the following:

1) A discount coupon for 25% discount on all books on our site

2) Occasional Notices of new books available

3) Occasional Email on other offerings of ours (Monthly)

Go to this link to sign-up:

http://personal-longevity.com/mkebooks/emailsignup/

And click this link to get the FREE 102 page Ebook titled "Secrets of Many Things"

If you have any questions about this book or other subjects please contact the Author at:

mke@mkettingtonbooks.com

Crazy, Strange, and Disruptive Ideas

Crazy, Strange, and Disruptive Ideas

Table of Contents

1.0   Introduction 1
2.0 What is Insane and What is just Strange?....................3
3.0 Some Far Out Ideas 7
   3.1 Thoughts turning Into Insects.....................................9
   3.2 Can Thoughts Create Reality?.................................13
   3.3 Duplicates of Each of Us..17
   3.4 Reptilian and Other Aliens 19
   3.5 Is Time Travel Possible?...27
   3.6 The Hollow Earth 33
   3.7 Is the Universe a Hologram? 37
   3.8 Everyone has Superhuman Abilities 41
   3.9 The Singularity is a Delusion 45
   3.10 Meeting Alien Spirits 47
   3.11 How to Clear Haunted Houses 51
   3.12 The Probability of the Future..................................55
   3.13 Individual Paths & Dimensions 59
   3.14 Life in Earth's Interior 61
   3.15 Are the Stars Alive? 65
   3.16 Longevity and Immortality 67
4.0 Disruptive Ideas...75
   4.1 Tectonic Plates: 77
   4.2 The Internet & Smartphones....................................81
   4.3 Oil and Gas Fracking 85
5.0 Potential Paradigm Disruptions 87
   5.1 Making Everyone Tiny 89
   5.2 CryptoCurrency Revolutions 91
   5.3 Direct Voting 93
   5.4 All Universities Online 95
6.0 Summary 99
Bibliography..101

Crazy, Strange, and Disruptive Ideas

Crazy, Strange, and Disruptive Ideas

1.0 Introduction

My focus has always been on understanding the real world. At the age of ten I read a book called "Stranger Than Science" By Frank Edwards which really opened my eyes and my mind.

This book was written in 1959 and as a child growing up in the nineteen sixties I was bombarded with unusual things in the news.

My dad was also an Engineer and my mother a Teacher so this background and their influence gave me an appreciation for the real world. As a result, I became and\ Engineer in the nineteen seventies and also a student of the unusual and paranormal at that time.

Was involved in paranormal and spiritual development classes in college as well as paranormal investigations. (Some fascinating stories of what occurred—these are for another time)

My life experiences and research have led me to become an "Out of the Box" thinker as a result of experiencing many strange and unusual phenomena.

Having written over sixty books on the paranormal, spirituality, aliens, Atlantis , and other unusual topics, the thought struck me to delve into the most far out concepts and ideas around.

Crazy, Strange, and Disruptive Ideas

This is the purpose of this book. To explore the most far out and even insane statements about our world, reality, and the universe.

It has been said that genius and insanity are closely related. I hope you enjoy this exploration and don't go insane in the process. LOL

Some thoughts which people think are stupid and insane today-later become disruptive ideas which change the world. Keep an open mind about the world—you will be pleasantly surprised with what you learn.

Crazy, Strange, and Disruptive Ideas

2.0 What is Insane and what is just Strange?

What is insanity? The definition has changed over history and with cultures.

In the old Soviet Union, you might be considered insane and sent to an institution if you were politically out of step with the leadership of the country and the local common beliefs about communism. (Or at least what people would voice in public.) Under this definition, most citizens of the United States who love freedom would be in institutions.

What about primitive Indian tribes in the Amazon? Most of these people would consider it normal that they commune with spirits in the Jungle and the spirits of ancestors. In our modern western society you would be considered pretty weird at a minimum and possibly subject to being put into a mental institution if the condition persisted.

In ancient Rome it was considered fun and recreational to watch Gladiators in the Colosseum fight and kill each

Crazy, Strange, and Disruptive Ideas

other; or to watch the massacres of Christians by wild animals. In today's society watching this type of sport would be considered totally disgusting and possibly illegal.

These are just a few examples of the norms of society for how persons should act can be totally different depending on cultures and timeframes.

Most psychiatrists today in the United States subscribe to the belief that you can believe in whatever you want as long as you aren't a danger to yourself or others. This is a healthier approach to the subject of insanity and the need for institutionalization.

However, there are many ideas which are still considered "wacko" and believers in those ideas considered pretty "mentally ill".

I would classify beliefs in concepts which don't make any logical sense as "wacko". You can propound almost any belief, but if you don't have any evidence to back it up, then you are going out on a limb which can easily be cutoff.

My strong beliefs in Prophecy might be considered wacko, but having had vivid prophetic experiences and seen them come true, I'm very well convinced of their validity. Of course I would have a lot of problems convincing others that what I experienced is true.

To me then, it isn't important if something sounds insane. What is important is does the statement, or idea, or concept possibly contribute anything to our understanding of ourselves or the universe.

Crazy, Strange, and Disruptive Ideas

Also, can you handle your experiences without becoming a danger to yourself or others? Some paranormal experiences can cause people to have nervous breakdowns. At the least these experiences can cause a lot of uncertainty the events we might have thought could happen—have really happened.

Crazy, Strange, and Disruptive Ideas

Crazy, Strange, and Disruptive Ideas

3.0 Some Far Out Ideas

As part of the theme of this book I want to present you with some really far out ideas. So please open your mind.

Some of these ideas might seem insane, but I believe it's very healthy to explore all sort of "out of the box" concepts to expand our minds and maybe come up with some unique solutions to today's problems and challenges.

You don't need to believe that these ideas are real or not. Just try to consider if these things might be possible. And what are the implications for our lives if these strange ideas are really true?

Crazy, Strange, and Disruptive Ideas

Crazy, Strange, and Disruptive Ideas

3.1 Thoughts turning Into Insects

Some ideas are so strange they are considered impossible or insane....

At University in Rensselear Polytechnic Institute in 1973 I took a beginning psychology course as an elective. In this course we saw a movie about insane people. It was really fascinating.

The one thing I remember vividly was a man who made the statement "My thoughts are turning into Insects". I tried for years to visualize how that could be. There must have been some logic for him to make that statement.

I finally decided that since many of our thoughts can become thought forms. This man might have been thinking about how the visualization of this thoughts might start transforming into images of insects.

Since I was also going through spiritual development at the time and experiencing many paranormal abilities like

Crazy, Strange, and Disruptive Ideas

premonitions, that maybe people in institutions are also experiencing these phenomena-but can't manage them.

It is certainly true that many persons in mental institutions are there to protect themselves or others from them. But many of these persons might have gotten that way because they couldn't handle their experiences with psychic phenomena experiences.

I know that my experiences with psychometry, taking in vital forces, and premonitions resulted in experiences which most people would consider impossible or insane. It can also be psychologically destabilizing to experience something which science and our western civilization says is impossible and know that the experiences are real.

These experiences made me question myself even though I'm normally an extremely stable person. I saw other people who experienced psychic phenomena and had nervous breakdowns as a result. They were also expecting what happened and supposedly prepared for these experiences.

Some persons probably have had paranormal experiences which they didn't know anything about and were totally unprepared for.

My mother is a good example. When my sister and I were small kids in Painted Post, New York in the early nineteen sixties my mother was gone for several weeks. When I asked my dad and grandparents they said she was sick and was in the Hospital for a while.

Crazy, Strange, and Disruptive Ideas

It wasn't until I was in my fifties and had experienced lots of spiritual and psychic experiences that I had a conversation with my Mom as to what happened to her.

It turns out she started seeing bright auras around people and thought she was going nuts. She talked to her minister who didn't know anything about this and couldn't help her.

So she checked herself into a hospital for a few weeks because she thought she was having a nervous breakdown. Her seeing of auras finally went away and she thought she was normal again. I had to explain to her that what she experienced was perfectly normal for a sensitive person to develop because she still didn't know anything about auras.

How many persons have had similar experiences, but couldn't handle what happened to them and had nervous breakdowns as a result? Many I suspect. The civilization and culture we live in affects how these experiences are accepted.

In India in the nineteen sixties a person would be accepted as having surely having had supernormal experiences, whereas in the United States at that time they would have been labelled nuts or crazy and had a good chance to end up in a mental institution.

Crazy, Strange, and Disruptive Ideas

Crazy, Strange, and Disruptive Ideas

3.2 Can Thoughts Create Reality?

When I was in college I went through psychic development training from learning to meditate and different exercises. One of the things I learned was about thought forms and visualization. The concept that our thoughts can create reality.

While this is something poo pooed by most people in our western culture, there are many spiritual gurus and even physicists who believe the same thing.

The physicists in question believe that our universe is really a hologram from the underlying reality and that what we see as three dimensional reality is really an illusion. That what we perceive as a three dimensional reality is really not there. That our minds build this reality based on misperceptions of the truth.

Crazy, Strange, and Disruptive Ideas

Something else I learned at that time from my middle aged, blind, psychic, mentor was that the realities we can imagine also achieve a kind of physical reality after we visualize them often enough.

This leads me to think about the fallout shelter I built in my backyard in 2002. At that time I lived on a hill near Los Angeles harbor. It was the year after the 911 terror attacks on the United States and I was worried that some foreign power might blow up a freighter in the harbor with a nuclear weapon. Having been trained in radiation exposure I knew that this could blow fatal radioactive particles over where I lived to kill people.

Never having built a fallout shelter before, I decided to visualize it before I started. I continued to do this every day for over a month. The size, the depth, the layout, how I would protect people inside of it from radiation. After a month it seemed solid in my mind and that I didn't need to do anymore visualizations.

Then I hired some workers to have them dig the pit in the hillside behind our home, pour concrete, put in steel reinforcements, and cover the whole thing back up to a depth of four feet of over burden. After that I added water tanks, a dehumidifier, electric power, and filtered air inlets. Food and other supplies were also added and the place was ready for usage.

After two months, the entire project was finished without a problem and at a low relative cost. I hadn't known if I could really build this and was worried it might be extremely expensive. The visualization was a large aid to help me build it. It is not a proof of visualizing reality which becomes reality. However, to me it was proof because of the

Crazy, Strange, and Disruptive Ideas

confidence I felt after the visualization exercise was done and my psychic certainty that it would happen.

Other thoughts of mine about visualization include the idea that television and movies create their own realities. I'm not the first to think of this since this theme has been used many times in shows like "The Twilight Zone" and "The Outer Limits".

What about the extended concept that a character visualized in a television show could also visualize their own newly created reality?

The Theosophists in the nineteenth century believed in Thoughtforms and wrote books on the subject. One of the best known is "Thought-Forms" by C. W. Leadbeater and Annie Wood Besant

I'm one of those persons who subscribes to the idea that visualization of a goal can create reality. Whether the effects are because these thoughts bring what we want in the Universe closer, or just that it improves our attitudes and positive outlook for the event in question doesn't matter. What matters is how we can achieve our goals and objectives in the best way possible.

So I do recommend that you get yourself a visualization board and paste pictures on it about what the goals and results are that you want to achieve. This process is very popular in some areas and I agree it is worth doing.

Crazy, Strange, and Disruptive Ideas

Crazy, Strange, and Disruptive Ideas

3.3 Duplicates of Each of Us

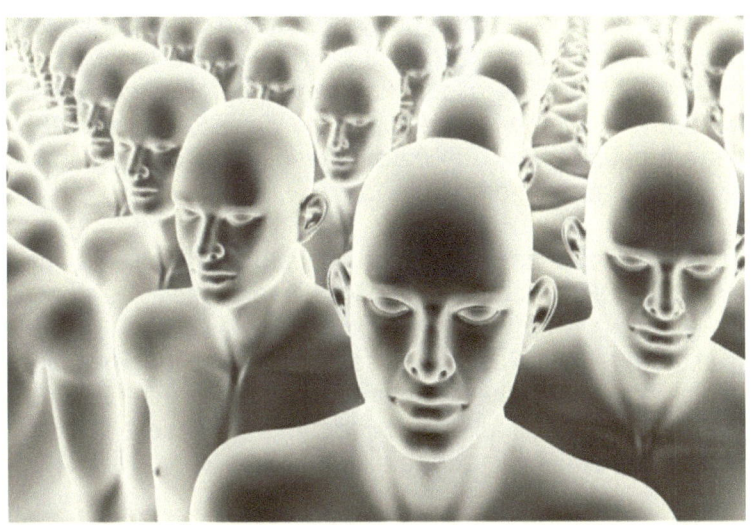

One of the thoughts I've had before is that there might be many other duplicates of myself throughout this Universe. Here is my logic:

Scientists estimate that there are over 2 trillion galaxies in the Universe.

We can calculate that this means there must be over 4,000,000,000,000,000,000,000 individual stars.

Now let's assume that only one out of a million stars has intelligent life on it. (This is probably low from what we know about planets around stars.) This still means there would be 4,000,000,000,000,000 stars with intelligent life. If only one millionth of those stars has human like life, and a couple of billion people per planet, we are still looking at 4,000,000,000,000,000,000 individual beings of human like life in the universe.

Crazy, Strange, and Disruptive Ideas

So how many combinations of genes are there which would be just like each of us in expression of our looks and minds? I would suggest one of a billion—but it is not an infinite number. There is no known number, but the overall numbers of human like beings in the Universe says that there must be many duplicates of each of us on planets going around different stars.

These numbers are really impossible to calculate with any scientific validity, but I intuitively believe this is to be true.

Sometimes I just close my eyes and imagine that my spirit is connecting with these other duplicates of myself around the universe. The feeling I get from this exercise is immensely satisfying and relaxing.

My thought is that these genetic duplicates of myself somewhere else in the Universe should be on the same wavelength as myself and therefore we are able to sense each other.

Feeling this affinity to others like myself gives me a feeling of confidence and peace going forward.

Crazy, Strange, and Disruptive Ideas

3.4 Reptilian and Other Aliens

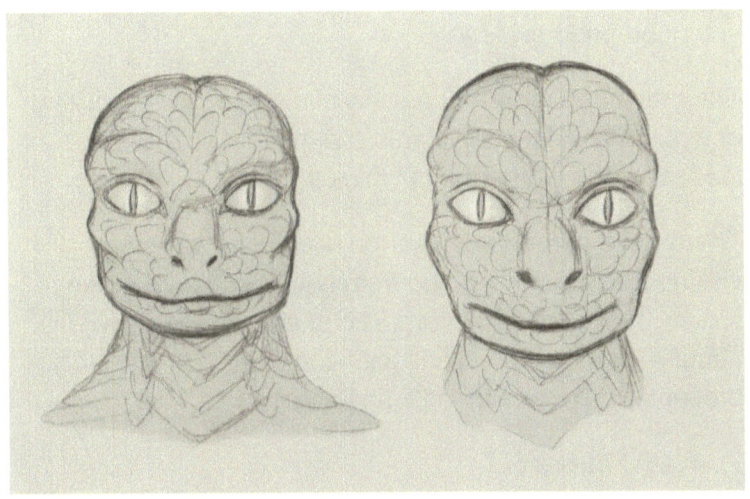

The idea that Reptilian Aliens exist hiding on our Earth has been popularized by David Icke. He insists that many famous persons around the world like the British Royal family and President George W. Bush are really aliens in disguise.

I have studied the subject of UFOs and Aliens and written a couple of books on the subject. Even I have a hard time in believing in Reptilian Aliens.

The whole idea is reminiscent of the Movie "Men in Black" that the world's most powerful people are aliens in disguise. This thinking is pretty far out there.

What would be the impact of this being true? It would mean that our society is run by beings who might have hostile plans for humanity.

Crazy, Strange, and Disruptive Ideas

The thing is that there is no evidence that these disguised reptilians exist. No pictures, no first person stories, nothing. So if you are going to propound such an incredible belief, you need good evidence.

Their might be reptilians as one of the Alien races visiting or who have visited the Earth, but them in disguise as leaders of the world? I don't think so.

There have purportedly been studies done on Aliens visiting Earth including one from NATO in the nineteen sixties. The story of the Top Secret study done by NATO about Aliens in 1964 is a fascinating story. The full story is presented here as I found it (in italics):

NATO Meets E.T.

Name: Robert O. Dean, retired Army command sergeant major

Claim: Back in the Sixties, NATO issued a classified report stating that UFOs were real, of extraterrestrial origin, and had visited the earth. This extraordinary report was said to come out of NATO's command center, the Supreme Headquarters Allied Powers, Europe (SHAPE), located then just outside of Paris, France.

Background: Dean, a highly decorated veteran, served on the front lines in both Korea and Vietnam. In 1963, while assigned to the Supreme Headquarters Operations Center (SHOC), SHAPE's war room, headed up by then-supreme allied commander of Europe, Gen. Lyman Lemnitzer, Dean claims he was

Crazy, Strange, and Disruptive Ideas

able to read the detailed 12-inch-thick NATO report on UFOs.

The Story: "SHAPE was one of those choice assignments. You had to have a spotless record and pass security background checks. I applied on a whim and got it. I was very proud and pleased. At SHAPE, I was put through more security checks, given a Cosmic Top Secret (yes, this is a real term) clearance, the highest NATO has, and assigned to the Supreme Headquarters Operations Center, known as SHOC, the NATO war room. In those days, the activity would run hot and cold and much of it would depend on how the Soviets wanted to play it. The most intriguing thing to me was that we were continually having a problem with large, metallic, circular objects that would appear over central Europe; these were reported as visual phenomena by our pilots and appeared on radar as well. Some flew in formation, and most of the time we spotted them coming out of the Soviet Union, over East Germany, West Germany, France, and then they would often circle somewhere over the English Channel and head north, disappearing from NATO radar over the Norwegian Sea. These objects were very large, moving very fast, at very high altitudes--higher than we could reach at the time--and they seemed obviously under intelligent control

"I was told this had been going on for some time and that in February 1961 there had been quite a scare. Fifty of these objects were spotted on radar and headed in formation from the Soviet Union toward Europe, flying at about 100,000 feet. The Soviets had closed all borders. Everybody went to red alert. All hell broke loose. We really thought `The War' had started. We scrambled. We knew the Russians were

Crazy, Strange, and Disruptive Ideas

scrambling. It was the largest number of these objects that had been seen. Fortunately--and only by the grace of God--we didn't start bombing and neither did the Russians. In nine minutes, they were gone.

"I was told that then-Deputy Supreme Allied Commander of Europe, Sir Thomas Pike, had been repeatedly requesting information from London and Washington about these objects, but nothing would ever come. We found out later that the Columbine-Topaz spy ring in Paris was intercepting everything and forwarding it to the KGB, which often got intelligence information even before we did. So Pike decided, I was told, to develop an in-house study to determine whether these objects were a military threat.

"In the meantime, the UFO matter literally brought about the establishment of direct communication between the East and West in 1962, which I have always found interesting and ironic. We had pretty well determined by that time that these were not Russian craft, and the Russians had determined they were not ours. So, we came to an understanding, and a direct telephone line was opened between SHOC and the Warsaw Pact Headquarters Command. Of course, a setup was always a possibility, so we had backup ways of checking out whether the Russians were being truthful. But since we were both armed to the teeth and World War III was just ticking away, it was a logical step in the right direction. That idea developed into the hot-line between the president of the United States and the soviet premier, following the Cuban Missile Crisis.

"Well, by the time I arrived in 1963, everybody had been talking about the study, and I had heard the rumors, seen the blips on radar, witnessed the

Crazy, Strange, and Disruptive Ideas

commotions, and some of us occasionally even talked about the possibilities. But nothing really prepared me for what I started to read in the early morning hours one night in January 1964.

"It was about 2:00 a.m. and a relatively quiet night when the SHOC controller on duty went into the vault and came out with this huge document. `Take a look at this,' he said. The title was simply Assessment: An Evaluation of a Possible Military Threat to Allied Forces in Europe. It was numbered, #3, stamped Cosmic Top Secret, had eight inches worth of appendices, dozens of photographs, and had been signed into the vault by German colonel Heinz Berger, SHOC's head of security. I quickly learned that it was based on two and a half years of research, was funded by NATO money, and that only 15 copies were published--in English, German, and French. Each one was numbered. All were classified and ordered to be kept under lock and key.

"Every time I got the chance, from then until I left, I would read a section or two in it. It was the most intriguing document I'd ever read. It was put together by military representatives of every NATO nation and also included contributions from some of the greatest scientific minds. These objects were violating all of our known laws of physics, and the study team had gone to Cambridge, Oxford, the Sorbonne, MIT, and other major universities for input on chemistry, physics, atmospheric physics, biology, history, psychology, and even theology, all of which were separate appendices.

"I read about theories on Einstein's sought-after unified-field theory, the high radiation at various landing sites, and UFO reports that dated back to the Roman

Crazy, Strange, and Disruptive Ideas

ea. and up to our own F105 pilots' sightings and encounters, and on and on. I had always been a skeptic, but this report, well...it concluded that this stuff was not science fiction.

"I read about contact encounters. One incident that had just happened in 1963 involved a landing on a Danish farm. According to the report, the farmer went aboard with the two little beings and two more human-looking men who spoke to him in Danish. The report included parts of his interrogation by government authorities and their conclusions that he was telling the truth. In another incident, according to the reports, a craft landed on an Italian airfield and offered to take an Italian sergeant for a ride. He wet his pants--that's what it said--and was so scared, he didn't go.

"The appendix that really got to me was titled `Autopsies.' I saw pictures of a 30-meter disc that had crashed in Timmensdorfer, Germany, near the Baltic Sea in 1961. The British Army, according to the report, got there first and put up a perimeter. The craft had landed in very soft, loamy soil near the Russian border and so hadn't destructed, but one-third of it was buried in. We and the Russians, who also quickly showed up, had both tracked it.

"Inside, there were 12 small bodies, all dead. There were pictures of the bodies, which looked like the beings known as the `grays,' being laid out and then put on stretchers and loaded into jeeps, and autopsy photos, too. Some of the little grays appeared to not be a reproductive-capable species. The autopsy guys concluded, according to the report, that it looked as if they had been cut out of a cookie cutter--clones with no alimentary tract. They did not ingest or process food as

Crazy, Strange, and Disruptive Ideas

we know it, nor did it appear that they had any system for elimination.

"The craft itself was cut up like a pie into six pieces, put on lowboys and hauled off. Scuttlebutt was that it was given to the Americans and flown to Wright-Patterson Air Force base in Ohio. I looked at these pictures and couldn't believe it. My skin got cold and I thought, My God. I had never really believed we were all alone in the universe, but this was hard to swallow.

"The major conclusions in the NATO report blew me away. There were five:

1) The planet and human race had been the subject of a detailed survey of some kind by several different extraterrestrial civilizations, four of which they had identified visually. One race looked almost indistinguishable from us. Another resembled humans in height, stature, and structure, but with a very gray, pasty skin tone. The third race is now popularly known as the grays, and the fourth was described as reptilian, with vertical pupils and lizard like skin.

2) These alien visitations had been going on for a very long time, at least 200 years--perhaps longer.

3) The extraterrestrials did not appear hostile since if that were their intent they would have already demonstrated their malevolence.

4) UFO appearances and quick disappearances as well as the flybys were demonstrations conducted on purpose to show us some of their capabilities.

5) A process or program of some sort seemed to be

Crazy, Strange, and Disruptive Ideas

underway since flybys progressed to landings and eventually contact.

"I wanted so badly to copy this thing. I did take a photograph of the cover sheet, which wasn't in and of itself classified. But I didn't want to wind up in Fort Leavenworth. So instead I would go to the bathroom and take notes--surreptitiously, very carefully.

Crazy, Strange, and Disruptive Ideas

3.5 Is Time Travel Possible?

Time Travel is a favorite subject of Science Fiction books and there are many thought provoking stories in this genre. I've read lots of those stories and really enjoy the time twisting ideas. Some physicists think time travel is possible.

Here is a quote about Steven Hawking the late great physicist:

> *Even the world-renowned physicist Stephen Hawking was entranced by the idea of time travel before his death this year, when he discussed in the Daily Mail how a black hole could make it possible. "Around and around they'd go, experiencing just half the time of*

Crazy, Strange, and Disruptive Ideas

everyone far away from the black hole. The ship and its crew would be traveling through time," he wrote in 2010.

I also think our consciousness can experience other times. Read this story I extracted from my book "God Like Powers and Abilities" and think about what might be possible. Here is a story which indicates that some type of time travel exists. Consciousness only or bodies too I don't know.

Visiting Versailles Palace in France in the past:

The classic of time slip tales occurred in August 1901, when two Englishwomen on holiday, Annie Moberly, Principal of St. Hugh's College in Oxford and Dr. Eleanor Frances Jourdain, visited Paris. After a short stay in the capital, they went on to Versailles.

After visiting the palace they began searching for the Petit Trianon but became lost. As they wandered the grounds, both women began to feel strange, as if a heavy mood was oppressing their spirits.

Two men dressed in "long grayish-green coats with small three-cornered hats" suddenly appeared and directed the women to the Petit Trianon.

They strolled up to an isolated cottage where a woman and a 12- or 13-year-old girl were standing at the doorway, both wearing white kerchiefs fastened under their bodices.

The woman was standing at the top of the steps, holding a jug and leaning slightly forwards, while the

Crazy, Strange, and Disruptive Ideas

girl stood beneath her, looking up at her and stretching out her empty hands.

"She might have been just going to take the jug or have just given it up I remember that both seemed to pause for an instant, as in a motion picture," Dr. Jourdain would later write.

The two Oxford ladies went on their way and soon reached a pavilion that stood in the middle of an enclosure.

The place had an unusual air about it and the atmosphere was depressing and unpleasant. A man was sitting outside the pavilion, his face repulsively disfigured by smallpox, wearing a coat and a straw hat.

He seemed not to notice the two women; at any rate, he paid no attention to them.

The Englishwomen walked on in silence and after a while reached a small country house with shuttered windows and terraces on either side. A lady was sitting on the lawn with her back to the house.

She held a large sheet of paper or cardboard in her hand and seemed to be working at or looking at a drawing.

She wore a summer dress with a long bodice and a very full, apparently short skirt, which was extremely unusual. She had a pale green fichu or kerchief draped around her shoulders, and a large white hat covered

Crazy, Strange, and Disruptive Ideas

her fair hair.

At the end of the terraces was a second house. As the two women drew near, a door suddenly flew open and slammed shut again. A young man with the demeanor of a servant, but not wearing livery, came out.

As the two Englishwomen thought they had trespassed on private property, they followed the man toward the Petit Trianon.

Quite unexpectedly, from one moment to the next, they found themselves in the middle of a crowd--apparently a wedding party--all dressed in the fashions of 1901.

On their return to England, Annie Moberly and Eleanor Jourdain discussed their trip and began to wonder about their experiences at the Petit Trianon.

The two began to wonder if they had somehow seen the ghost of Marie Antoinette, or rather, if they had somehow telepathically entered into one of the Queen's memories left behind in that location.

As if to confirm their suspicion, Moberly came across a picture of Marie Antoinette drawn by the artist Wertmüller.

To her astonishment it depicted the same sketching woman she had seen near the Petit Trianon. Even the clothes were the same.

Intrigued by the growing mystery, Jourdain returned to

Crazy, Strange, and Disruptive Ideas

Versailles in January 1902 and discovered that she was unable to retrace their earlier steps.

The grounds seemed mysteriously altered. She then learned that on October 5, 1789 Marie Antoinette had been sitting at the Petit Trianon when she first learned that a mob from Paris was marching towards the palace gates.

Jourdain and Moberly decided that Marie Antoinette's memory of this terrifying moment must have somehow lingered and persisted through the years, and it was into this memory that they had inadvertently stumbled.

Crazy, Strange, and Disruptive Ideas

Crazy, Strange, and Disruptive Ideas

3.6 The Hollow Earth

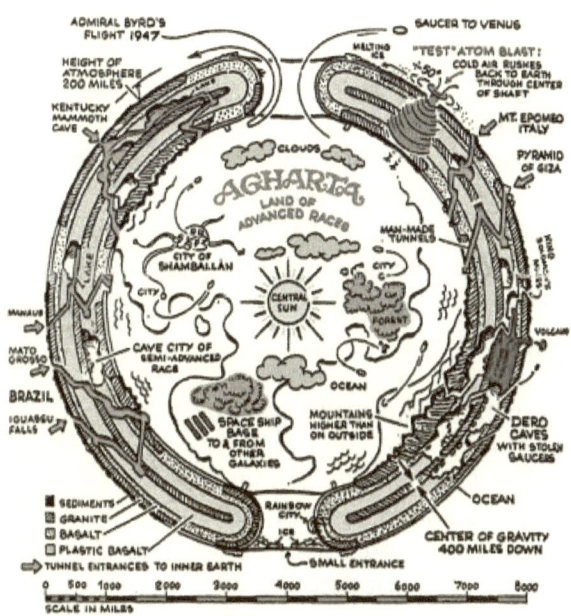

The belief in a Hollow Earth is really interesting and from many cultures. Quite a few Indian tribes said that their ancestors came from inside the Earth. Here are some of those beliefs:

- The Navajos believe that they emerged at a place known only to them in the Navajo Mountains.
- The Aztecs feel that they were one of seven tribes that came out of the caverns of Aztlan.
- The Zuni believe that, "in the old days all men lived in caves in the center of the earth".

Crazy, Strange, and Disruptive Ideas

- The elders of the Hopi people believe that an entrance in the Grand Canyon exists which leads to the underworld.
- And there is a tale about a tunnel in the San Carlos Apache Indian Reservation in Arizona which is said to lead inside the earth to a land inhabited by a mysterious tribe.
- According to the Pawnee story of creation "All living things came from under the ground".
- Chekilli, the head chief of the Creeks Indian tribes said "The earth opened in the west, where its mouth is. The earth opened and the Creeks came out".

Not just Indians talked about living underground:

- Ancient Egyptians had their invisible Underworld of the Dead, which was ruled over by Osiris.
- The Greeks had a dark Underworld of the Dead ruled over by Hades.
- The Norsemen, Teutons and Vikings had their Underworld, a place called Nilfheim - a cheerless world of ice and darkness that was their Land of the Dead.
- The Mayan underworld contained several deadly places, such as "The House of Gloom", "The House of Cold", "The House of the Jaguars" and "The House of Fire".
- According to Inca legends the first of the Inca race were four brothers and four sisters who were said to have emerged from a cavern close to Cusco in Peru.

The general modern belief is that there is some type of entrance to the underworld at the poles. The Shaver

Crazy, Strange, and Disruptive Ideas

Mysteries are one of the modern beliefs about the Hollow Earth:

> *The science fiction pulp magazine Amazing Stories promoted one such idea from 1945 to 1949 as "the Shaver Mystery". The magazine's editor, Ray Palmer, ran a series of stories by Richard Sharpe Shaver, claiming that a superior pre-historic race had built a honeycomb of caves in the Earth, and that their degenerate descendants, known as "Dero", live there still, using the fantastic machines abandoned by the ancient races to torment those of us living on the surface.*
>
> *As one characteristic of this torment, Shaver described "voices" that purportedly came from no explainable source. Thousands of readers wrote to affirm that they, too, had heard the fiendish voices from inside the Earth. The writer David Hatcher Childress authored Lost Continents and the Hollow Earth (1998) in which he reprinted the stories of Palmer and defended the Hollow Earth idea based on alleged tunnel systems beneath South America and Central Asia.*

So do these caverns and cities inside the Earth exist? I don't know, but I'd like to find the purported entrance to one of them and go exploring. Of course I don't know if I would be coming back so should settle my affairs first before setting out.

Crazy, Strange, and Disruptive Ideas

Crazy, Strange, and Disruptive Ideas

3.7 Is the Universe a Hologram?

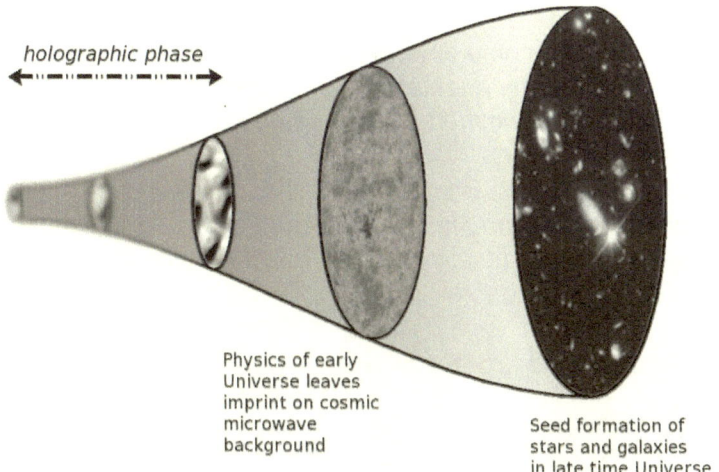

holographic phase

Physics of early Universe leaves imprint on cosmic microwave background

Seed formation of stars and galaxies in late time Universe

A Hologram that we can view is a three dimensional picture produced by lasers and was first invented in the 1960s. It is by definition not a real physical object although it looks like one.

Now some physicists believe that our Universe may be a hologram above the basic substrate or foundation of our Universe. A science article gives you a little bit of the flavor of what theoretical physicists are dealing with:

> *So how do you get a hologram?*
>
> *Let's go back to that two-dimensional surface covered with entangled qubits. Since the value of a qubit changes depending on the value of its entangled pair, there's a degree of indeterminacy built into the system.*

Crazy, Strange, and Disruptive Ideas

If you haven't yet measured the first qubit, you can't be sure about the second. The amount of uncertainty in any given system is called its entropy.

As qubits become entangled and disentangled, the level of entropy rises and falls. You wind up with fields of entropy in a constantly changing state.

The holographic principle holds that our three-dimensional world is a representation or projection of all this activity taking place on a two-dimensional surface full of qubits.

Putting it all together

It's always bothered physicists that there is one set of rules for the microcosmic, quantum mechanics, and another for the macrocosmic, the theory of relativity. It doesn't make sense that there should be two different and incompatible groups of mathematical formulas at work in our universe. Physicists assume there must be some way to bring them into harmony.

So therein lies the central question for Headrick and his colleagues: Starting in the two-dimensional realm of qubits and quantum mechanics and then scaling up in size, how precisely do we wind up with bits and relativity? It's a matter of constructing a single mathematical model that explains the transformation.

Figure it out and you'll have solved one of the biggest mysteries in theoretical physics. From the tiniest to the largest phenomenon, we'll have a unified theory of reality.

Crazy, Strange, and Disruptive Ideas

Right now the holographic principle remains an unproven theory. Where it will lead next is an open question. Odds are though, it'll be stranger than anything yet imagined in science fiction.

The impact of all this is that the physical reality we see is all an illusion. We know from current science that this is true since the spaces between atoms is huge so much of what we perceive as solid is really empty space.

Knowing this is true and that physicists still have many questions about our physical reality should let us consider that many psychic powers like teleportation or psychically manipulating our environment might have a basis in our real reality.

Crazy, Strange, and Disruptive Ideas

Crazy, Strange, and Disruptive Ideas

3.8 Everyone has Superhuman Abilities

The idea that everyone has superhuman abilities sounds like a Marvel Comics idea and makes for great plots in superhero movies, but what if it's true?

I approach this subject as a student of spiritual growth, experiencer of many psychic abilities, and writer of a number of books about the paranormal.

There is a book written in Sanskrit and translated to English titled "The Yoga Sutras of Patanjali". It was written over 2,000 years ago in India and describes the spiritual development process from the perspective of those persons who had experienced this development in Indian cultures at the time. I recommend reading this book to get a good feel about how the ancients viewed spiritual development and the side effect abilities which students of

Crazy, Strange, and Disruptive Ideas

spiritual development might experience. The book is available for free online and can be purchased too.

Spiritual development in this context is defined as learning to meditate and other processes to help oneself center themselves spiritually.

My book "God Like Powers & Abilities" discusses many of the abilities one can achieve through the spiritual development process.

I can testify that through meditation and different exercises I learned how to take vital forces in through my chakras for healing, and that the new "openness" this produced in me is probably responsible for my wide ranging premonitions. See my book "Prophecy: A History and How to Guide" for the details.

So following the old idea "Truth is Stranger than Fiction" we should understand that our potential is much greater than anything we can conceive. Here is a partial list of abilities and experiences I've researched which I think do have a basis in fact and some persons have experienced:

- Telepathy
- Clairvoyance
- Seeing Auras
- Astral Projection
- Incredible Body Heat
- Psychokinesis-The Mental Control of Matter
- Psychic or Spiritual Healing
- Prophecy
- Great Powers of Magic
- Spiritual Possession

Crazy, Strange, and Disruptive Ideas

- Levitation
- Invisibility
- On Becoming Invulnerable
- The Microscopic & Macroscopic
- Cessation of Hunger and Thirst
- Teleportation
- Bi Location
- Infinite Knowledge
- Time Travel
- Immortality
- Creating or Changing Reality
- Spiritual Enlightenment

Crazy, Strange, and Disruptive Ideas

Crazy, Strange, and Disruptive Ideas

3.9 The Singularity is a Delusion

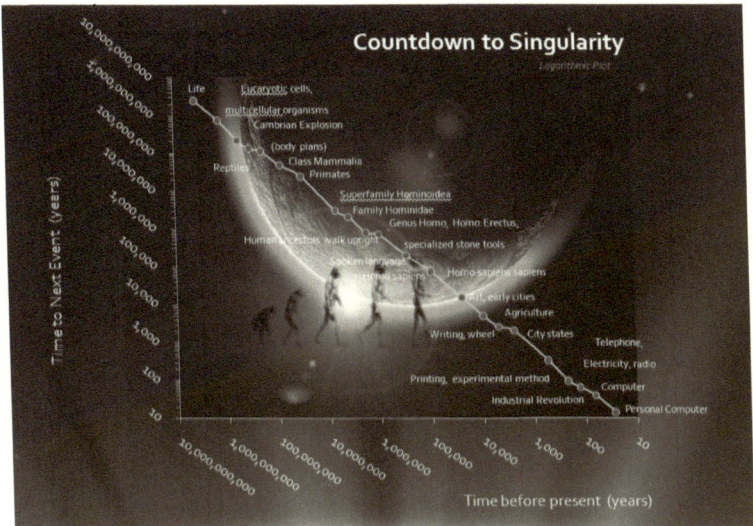

Many persons in the high tech world worry about artificial intelligence going rogue and taking over the world. They are concerned that artificial intelligence programs will achieve real consciousness and then use their knowledge and processing speed to take over and control the world. That it might even decide to get rid of humanity. This whole potential sequence of events is called the "Singularity".

This potential threat is governed by the idea that there would be no real difference between a machine intelligence and human intelligence. What the people who propose the Singularity don't understand that humans have more than just a physical intelligence. We also have a spirit which is in our bodies and connects to God.

Crazy, Strange, and Disruptive Ideas

This view is not just a Christian view, but is one that most religions agree with. Therefore a machine intelligence would never really be alive, since it might simulate intelligence, but it would never have this spirit as part of its makeup.

My agreement with this view of spirituality is based on several of my observations about the Universe. First, I know from experience that premonitions are real. The only way for this to work is for our beings to have some view outside of the time stream. This would not be possible if we were purely logical beings. We must have some connection to a core part of our spirit which lives outside of time and space.

My second point is that many theorists believe that the three dimensional world we see is really a type of Hologram which is a more fundamental reality. If this is true, then again, there is much more to the Universe than just physical intelligences.

Scientists today don't know as much as they think they do about what makes up a person and the components needed to makeup and independent and thinking intelligence.

Crazy, Strange, and Disruptive Ideas

3.10 Meeting Alien Spirits

One popular question about Aliens is why haven't we met any? Most persons would say that that the distances to other stars might be too difficult to ever overcome. They might be right.

What about meeting them in another dimension? People who have out of body experiences (OBE) sometimes claim experiences of meeting aliens from other worlds in their OBE states.

This would actually make a kind of sense since any advanced alien species is also likely to be more advanced than humanity in a spiritual sense. And it is known that you have to be able to go deep into a meditative state and

Crazy, Strange, and Disruptive Ideas

have a somewhat spiritually oriented personality to achieve a conscious OBE.

Since communication on the astral plane is by telepathy and not voice, it might also be a much better way to communicate with aliens.

So let's start an Astral Space Program to push our efforts to meet aliens the easier and faster way!

Below is a guy named David McCready who has written an article about traveling the stars using astral projection. This is a snippet of his article:

Advanced alien races have been visiting Earth for millennium, but until very recently, they have revealed very little about themselves. However, this is all changing and alien contact of various descriptions is becoming commonplace.

> *I am known as David McCready, the author of The Great Simulator series of books which details how the illusion of life we experience is engineered. My specialist field of expertise is enabling people to personally observe how the illusion of our adopted human manifestation is created, along with how to easily transcend that illusion and regain full access to the Astral World.*
>
> *Once you see beyond the illusion of human identity, it becomes obvious that the human personality any of us experience being is just a temporary manifestation. Similarly, it becomes quite clear that we have all experienced life on other worlds. Furthermore, there is*

Crazy, Strange, and Disruptive Ideas

fundamentally only one of us, but the illusion makes it appear that there are instead many of us.

Inception

In early 2014 a research team from a group of higher dimension advanced aliens made contact with me. They conducted a two-way pairing operation in which their team leader initiated a portal into the Earth world, whilst I gained an equivalent portal into theirs. Their team leader indicated that it was convenient to identify him as the 'Doc', since his role was associated with maintaining the health of their space-time astronauts.

Advanced Astral Projection

Two way portals are essentially an extension of astral projection or controlled out of body exploration, by which beings from distant worlds are able to explore each other's planets.

A human, or alien, physical body is essentially a focusing machine designed to give you the experience of living in an illusory physical world such as the Earth. If you experience astral projection during sleep, your consciousness normally remains contained within the spirit component of that focusing machine. Hence exploring alien worlds is extremely difficult, as your spirit focusing machine tends to keep you earthbound.

The significant advance is to leave your earthly focusing machine (physical and spirit bodies) where it is, and instead share an alien focusing machine in its world. Once a pairing is initiated, the human consciousness experiences an instant astral link into that alien world. Similarly, alien researchers are able to explore the human world without the need to physically travel to it. This exchange is best conducted with all parties fully awake, as opposed to say, lucid dreaming, etc.

There is no limit on the number of pairings that can be put in place. An analogy is how many video skype addresses or phone numbers can you possess?

<u>Ending the Isolation</u>

Human beings have hitherto been developed in relative isolation with respect to enjoying easy communications with the rest of the universe. It also must be observed that the souls who inhabit humans or aliens are substantially similar, regularly changing resident bodies and physical home worlds. So up till now, the Earth was an environment in which you could experience relative isolation from your soul friends and families living elsewhere.

Since human culture still retains ingrained warlike tendencies, this is the sort of visitor most advanced alien species are cautious about physically entertaining in their home worlds.

The stepping-stone forward is to allow souls experiencing human form to have astral access to advanced alien worlds. This enables human minds to become acquainted with how more advanced societies operate.

In effect, human beings are being given a sort of "inter-galactic internet" connection which is enabling them access to a far bigger picture.

Crazy, Strange, and Disruptive Ideas

3.11 How to Clear Haunted Houses

I had an experience back in 1975 which changed my understanding of reality forever. Some friends and I went to capture a spirit from a haunted house. I present the story here as an example of what not to do with a haunted house. After the story I'll give you my thoughts for the better and safer way to clear a haunted house:

One evening in October a very scary event occurred which I would never forget. To be blunt, what happened scared the hell out of him and he never wanted to go through that experience again. It all started when Sam told him and another student Mark that one lady in the meditation group thought her house was haunted and wanted them to help clear it. Sam thought he could clear the house and wanted to make a paranormal experiment out of it.

Crazy, Strange, and Disruptive Ideas

On a Saturday night they took a compass, a camera with Infrared film, and a bottle of water—of a couple gallons-which Sam had treated psychically—saying that any entities found in the home would be drawn into it. Sam used a psychic technique he called polarization. The ideas behind the equipment was to use the film to shoot any hot or cold spots in the house, and the compass to see if any magnetic fields changed. The plastic bottle of water was to later be measured by Nuclear Magnetic Resonance back at the University to see what could be learned from how it had been changed. (NMR is now used daily in medical scanning equipment.)

Martin drove the three of them to an old home in Albany, New York where they met the woman who owned it and spent a couple of hours taking pictures and feeling psychically for hot and cold spots with their hands. Martin thought it was all interesting but maybe they were just on a wild goose chase. At one point Sam opened his bottle of water in the living room and said that he could sense different entities were being drawn into it. Martin thought that comment was ridiculous. After a couple of hours Sam declared the home was "Cleared" and they left. It was about midnight, but Sam wanted to stop to see his blind friend Carolyn. They took the bottle of water with them upstairs to Carolyn's apartment and had Pizza with her for another couple of hours. Martin gave the bottle of water to Sam who said he sensed lots of bubbles. Mark looked at the bottle and concurred. Then the three of them got back into Martin's car and started heading back to school around 2AM on the freeway next to the Hudson River. It was about a twenty mile drive.

Crazy, Strange, and Disruptive Ideas

Half way back the car's engine died. It was an old Opel Manta and the gas gauge didn't work too well. Martin got out and said "I need to flag down a car to get gas". He did that and quickly got a ride to the nearest gas station. Getting enough fuel to get home he got a ride back to the car which was parked on the side of the freeway. When he got there Sam and Mark were all upset. They said "Rocks started flying against the car and we know there is some type of entity in the car with us." Martin thought this behavior was paranoid and he said "It's late and we are all tired so just calm down." But as he started driving again he could feel there was something in the air in the back seat looking over his shoulder. Sam and Mark were in the backseat holding the water bottle at this point. Sam and Mark kept talking crazy and claimed the "thing" was attacking them. Martin told them to relax-they would be home soon. He drove onto the bridge over the Hudson River and back into Troy then up the hill to the school. There was a modest rain falling as he drove up the hill.

Near the top of the hill it happened: Martin felt a conscious and evil invisible cloud surrounding his head and psychic fingers penetrating his brain like a knife. It felt like his mind was being raped. He screamed, almost lost control of the car and struggled to regain control of his mind and the car. Sam and Mark were also yelling saying they were under attack too. He drove another mile on side streets then pulled over to the side of the road next to a storm drain. It was raining and lots of water was draining into it. Getting out and opening the rear door of the car he grabbed the bottle of water from his friends. Then he went to the storm drain where he emptied the bottle. After that he kicked the bottle into the drain.

Crazy, Strange, and Disruptive Ideas

They all calmed down as they want to Sam's apartment but still felt the "Entity" from the bottle was hovering near them. Sam called a priest friend to say a prayer for them and it was nearly dawn when Martin went back to his dormitory room. For months after that Martin could feel the "Entity" hover near him mentally until finally it went away. He never wanted to repeat anything like this experience again. This was definitely not a happy experience in his life. It was the most scared he had ever been. Martin seriously asked himself if getting involved with these types of things was a real path to enlightenment or just a way to get himself killed.

So now when a friend asks me how to clear a haunted house I have a totally different approach. Instead of trying to capture and eject spirits, I tell them to get a group of sensitive persons, Get them to relax into a meditative state, and then visualize a lot of spiritual energies to raise the vibrational level of the home. Think of love, and God, and just visualize these spirits being uncomfortable and leaving.

Now, if you ever have a home full of spirits, you know what to do!

Crazy, Strange, and Disruptive Ideas

3.12 The Probability of the Future

I don't know if this topic would be considered crazy—more likely it is considered unknowable by most people. In my book "Prophecy: A History and How to Guide" you can read more about my experiences with premonitions. These include being in a surfing accident, avoiding a plane crash, and seeing the 2004 Indian Ocean Tsunami multiple times up to a year before the event.

These experiences have led me to a better understanding of the probability of the future. In other words, the future is not fixed, but it has a certain amount of momentum

Crazy, Strange, and Disruptive Ideas

towards certain events occurring. This was illustrated to me when I avoided that plane crash.

Here is the story of how I avoided a plane crash:

> *During early August of 1998, my wife and I decided to send her and our kids to visit her mother in Barcelona, Spain. I was going to buy a ticket separately, and meet them there during early September.*
>
> *When I started to call the travel agent to book my ticket I had a terrible feeling of fear about taking the flight. I tried two other times to book the ticket during the week for a September 2nd departure, and each time I got the same strong feelings of fear and death.*
>
> *I have always prayed and tried to guard myself mentally to avoid disasters, so finally I took the warning seriously and decided not to go at all. This was very difficult to do since I really wanted to see my wife and kids, and this meant I would be home alone for a month.*
>
> *Work wasn't an excuse either, since I wasn't doing any really heavy contract work at the time and could easily have taken the time off. I called my wife and told her my decision, and she was surprised, but agreed for me to follow my instincts.*
>
> *On September 2nd the Swissair disaster occurred on a plane leaving Kennedy airport in New York, which crashed in Newfoundland Canada with all lives lost. I would not have originally been booked on that flight, but could have easily ended up on it since I was due to fly through Kennedy airport, and any delay might have*

Crazy, Strange, and Disruptive Ideas

caused me to switch planes. I will never know for sure, but this was a very strong warning.

I should also mention that for several years before this event I had strong feelings that my I would be killed in the near future. After this happened those feelings ended.

The above story showed me that we can change the future but it might take a lot of effort due to the momentum of events.

So what are the implications of this knowledge that future events are not fixed? To me this answers the age old question that we are creatures of free will—not a fixed destiny.

Some major events like the 911 attacks might be something that no one person can change. Those types of events would take hundreds if not thousands of persons to change them.

Also, this knowledge should give each of us a certain level of psychological freedom to pursue our interests and passions without feeling that our actions are futile in the greater scope of the world.

Crazy, Strange, and Disruptive Ideas

Crazy, Strange, and Disruptive Ideas

3.13 Individual Paths & Dimensions

A fascinating movie is "The Butterfly Effect" which has a main theme that our individual actions change the course of reality. In the movie, the guy whose life we are following can put himself back in time to experience life based on new decisions. He also feels the effects of those decisions in the present moment.

What if we carry this idea on step further? What if every decision we make splits future paths to make a new reality for each decision, and the alternatives to those decisions continue on their own paths.

Crazy, Strange, and Disruptive Ideas

Then we are creating an increasing and ultimately infinite number of realities based on all of our decisions. All those realities keep existing as the one we are in continues.

What if we can shift realities just by focusing our minds? If possible then we could experience the lives we would have led by making alternative decisions.

Switching back and forth between different realities would be quite the experience. I wonder if this occurs to everyone.

Have you ever misplaced something, looked everywhere, and then found it again in a place you already searched? If so maybe you have shifted to another reality.

Here is an interesting thought—what if when we die, we actually switch our bodies to another reality where we can go on living?

Is it possible that we shift realities many times during our lives and that maybe this is the cause of why we sometimes find things looking a little different than we remember them?

Or what about some new items where you say to yourself- "I thought that this news was about a different person or event" How does this happen?

Crazy, Strange, and Disruptive Ideas

3.14 Life in Earth's Interior

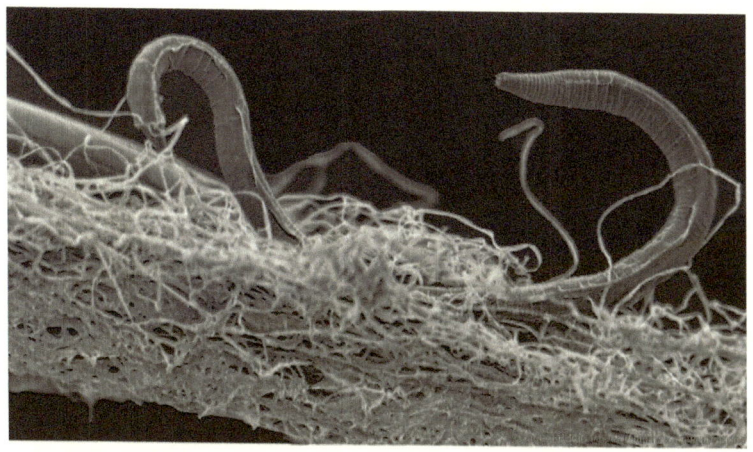

Where do we find life on Earth? On the surface naturally we would say. Maybe there are a few bats and fish in caves underground too. Well, one of the most exciting discoveries in recent decades is that the life beneath the surface of the Earth is incredibly large and deep.

The total biomass underground, down to the depth of several miles is probably much greater than the biomass of life living on the Earth.

Here are some portions of an article on the subject:

Indeed, studying Earth's deep microbial life has already pushed the understanding of the conditions under which life can thrive. Researchers have drilled miles into the seafloor and sampled the microbiomes from mines and boreholes at hundreds of sites around the world. Data from these sites suggest that the world's

Crazy, Strange, and Disruptive Ideas

deep biosphere spans roughly 500 million cubic miles (2.3 billion cubic kilometers) — about twice the volume of all the Earth's oceans — and houses about 70 percent of all the planet's bacteria and single-cell archaea.

Some of these species make their homes among the world's hottest, deepest niches. A frontrunner for Earth's hottest organism in nature is the single-celled Geogemma barossii, according to the statement. Living in hydrothermal vents on the seafloor, this microscopic spherical lifeform grows and replicates at 250 degrees Fahrenheit (121 degrees Celsius), well above the boiling point of water at 212 degrees F (100 degrees C).

Meanwhile, the record for deepest-known life so far is about 3 miles (5 km) below the continental subsurface and 6.5 miles (10.5 km) below the ocean's surface. Under this much water, extreme pressure becomes an unavoidable fact of life; at about 1,300 feet (400 meters) depth, the pressure is about 400 times greater than at sea level, the researchers wrote.

These recent discoveries have lots of interesting implications such as:

If there was a totally destructive Nuclear War or Disaster on Earth, the biomass below the Earth would probably repopulate the surface in only a few hundred to a few thousand years.

Crazy, Strange, and Disruptive Ideas

That life on other planets can exist in much more extreme conditions than we previously thought due to what Earths underground life can endure.

Temperatures of 250 degrees Fahrenheit or more, and extreme pressures like those found over six miles below the bottom of the Earth's Oceans are possible habitats for life. These are very extreme conditions of what we should consider possible to life to exist in.

Crazy, Strange, and Disruptive Ideas

Crazy, Strange, and Disruptive Ideas

3.15 Are the Stars Alive?

So we know that the Earth or Gaia is alive on the surface and deep into the Earth. Many spiritual people think that physical matter can be conscious and that larger groupings of matter have their own consciousness.

The Theosophists used to write about the Solar Logos, Earth Consciousness and more. Given that I believe that consciousness exists at all levels of reality, then this view makes a lot of sense.

I have sometimes sat next to a large tree, and went into a meditative state to commune with the tree. I often felt it was speaking back to me. Call it a delusion, but I do feel that every rock has consciousness.

Crazy, Strange, and Disruptive Ideas

So let's expand this idea. Many people believe that our Mother Earth or Gaia is conscious. It is often called other names too like "The Earth Mother". This consciousness is protective of all life on Earth.

Next would be the Solar Logos or consciousness of the Sun and the Solar System. Impossible to look at without heavy glasses. It is watching out for the welfare of everything in the Solar System with a viewpoint of billions of years.

Through the process of meditation and visualization you can align with these great beings.

Next is the Milky Way Galaxy of 200 billion stars. There is a huge black hole at the center of the Galaxy. Maybe this is the soul of the Galaxy. The Galaxy would again view time as having begun not too many billion years before it since it has stars going back to a few billion years after creation.

What about Galaxy clusters? What do they think about? Maybe they are thinking about how to best link and arrange the galaxies which compose them.

There are many levels to look at and visualized these greater beings which may offer intuitive insights into their structures and meanings.

Our Brains and a cluster of galaxies are now known to have similar complexities. Is this another indication that intelligent beings may exist at many levels of existence in the cosmos?

Crazy, Strange, and Disruptive Ideas

3.16 Longevity and Immortality

I almost forgot to add a chapter on Longevity and Immortality since I've studied and researched this subject for ten years and consider it all normal.

The truth is that what I consider normal on this subject would strain the bounds of credulity for almost everyone.

Here are some of the things I've learned and experienced while researching this subject:

- I found many records of people in their mid-one hundreds all the way up to 200 years or more and all over the world.
- I learned that we can improve our longevity by decades and developed that knowledge into what I call "The 10 Principles of Personal Longevity" (See later in this chapter for more information)
- Met a man who claimed he was over 2800 years old and he told me all about the community of Immortals in the world. Was the basis for my book "The Commentaries of Living Immortals"

Crazy, Strange, and Disruptive Ideas

- One book talks about the oldest man in the world who is over 1000 years old and lives in a monastery of the jungles of Burma.
- That there are many long lived cultures around the world with a large number of centenarians. See my book "The Diets and Lifestyles of Long Lived Peoples"

Below is a list and description of the 10 Principles which are the basis for my online multimedia training program for Longevity Coaches:

The 10 Principles of Personal Longevity are:

- The Reality of Long Lived People
- Defining Your Purpose in Life
- Enabling the Life Urge
- Your Spiritual Health
- Having Love in Your Heart
- Energy Body Health
- The Science of Longevity
- Physical Body Health
- Using your Intuition for Safety
- Implementation of these principles

What are the 10 Principles all about?

The Reality of Long Lived People

The first principle is where I provide lots of evidence of people who have lived well over the age of 120 years old to 150-180-200, and even a 256 year old man from China:

Crazy, Strange, and Disruptive Ideas

LI CHING-YUN: The Longest Lived person of record-256 Years (Source-The New York Times-May 6, 1933)

The Second Principle of Life Purpose

One of the things that occurred to me when I was putting the 10 principles together was that if one doesn't have a reason to live, or purpose in life--then what is the point?

This meant I had to add a very important step of how you can develop your own life purpose, or bring it up to date with your phase in life. Without reviewing your purpose-- then none of the rest of the principles matter.

Enabling the Life Urge

Crazy, Strange, and Disruptive Ideas

Have you ever realized how we are all programmed to expect to live through certain stages in life and then die? It's so common in our society that we don't think it odd that we expect to die at a certain age?

Have you ever heard radio ads saying "You are getting up in your sixties and seventies" so it's time to come out to our cemetery and buy a plot"

How ridiculous is this? And do you see how much our subconscious has been programmed towards death?

This principle is all about reprogramming ourselves to have a more positive outlook on life and its possibilities.

Having a Spiritual Connection in Your Life

Most of us innately understand that we have a spiritual

core in the center of our being. It is this spiritual core that we need to connect with to enable our physical health too.

It doesn't matter what religion you are. Regular meditation, deep prayer, or just walking in the woods helps you make and keep that connection in your life.

Having Love in Your Heart

One of the most important things I learned in the last five years was that Unconditional Love is a real and physical thing. It is a powerful energy force in life and not just a philosophical belief system.

I considered it so important that I added it as a separate principle of longevity.

True Unconditional Love is healing, embodies happiness, and is a powerful part of our vital forces.

Energy Body Health

Crazy, Strange, and Disruptive Ideas

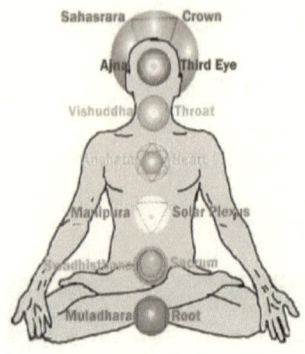

We all have an energy body which is part of our vital forces. The Indians talk about the "Chakras" and the Chinese talk about "Energy Meridians" in Acupuncture.

We should all learn different practices to keep our vital forces flowing for maximum health and vitality.

The Science of Longevity

Science and Medicine are making new discoveries all the time that we can take advantage of to extend our lives. Why not take advantage of these discoveries which provide new therapies and supplements to increase our

Crazy, Strange, and Disruptive Ideas

longevity.

There is also a lot we can learn from plants and animals. We all share the same genetic basis.

Some of these plants and animals live thousands of years and some cells are immortal.

What can we learn from them to apply to our lives?

Physical Body Health

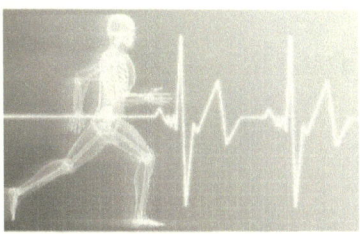

There are many types of supplements used for anti-aging for thousands of years. What can we learn about them that we can apply to our lives?

What other considerations about our physical health does nontraditional or alternative medicine offer?

Using Your Intuition for Safety

Crazy, Strange, and Disruptive Ideas

Once you have established your own long term health then what is the greatest danger you face?

ACCIDENTS

We can learn to use our intuition to make us safer as well as see potential future events which may be good too. Why not open up to the possibilities of how our spirit has this natural ability in all of us?

Implementing These Principles in Your Life

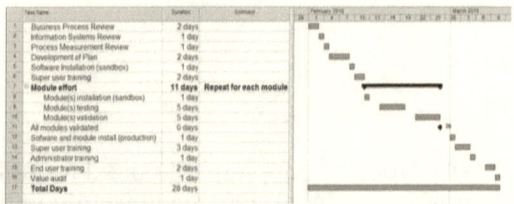

It's nice to read about all these concepts, but how can you really apply them to your own life?

This is what the chapter on implementation is all about, and it helps you plan a lifelong change in your health focus to live these principles and truly experience long term health, greater happiness, and extended longevity.

Crazy, Strange, and Disruptive Ideas

4.0 Disruptive Ideas

Disruptive ideas are those which are so different from the existing assumptions of the understanding of the world, or ways to accomplish something in such a new way that entire industries change or die.

Many disruptive ideas are considered insane or stupid when originally proposed, and after they occur seem obvious and the only way to do things.

Here are a couple of tables from my book "Future Predictions by an Engineer and Seer":

Far Out Possible Paradigm Shifts:

Focus	Paradigm Shift Type	Description	Prob	Time Frame
Gravity	Anti-Gravity	Use of Anti-gravity technology for transportation and in phases of society	80%	50 Years to 200 Years
Paranormal	Mass Psychic Abilities	That Psychic abilities will be generally accepted and people trained in their usage	95%	Now to 10,000 Years
Zero Point Energy	Zero Point	Mass Zero Point Energy production from the vacuum	10%	500 Years

Crazy, Strange, and Disruptive Ideas

Technology Driven Probable Paradigm Shifts:

Segment	Paradigm Shift Type	Description	Timeframe
Biology	CRISPR Technology	Large Scale Disease and Biology changes	Now-200 Years
Digital Currency	Crypto Currencies	Digital Currencies independent of any state	Now to 100 Years
Education	Remote Learning	A gradual shift towards most learning online	Now to 100 Years
Farming	Hydroponics	Large scale increased food production	Now-100 Years
Power	Fusion Power	Fusion Power and then miniaturization	30 years-300 years
Health & Medicine	Increasing Longevity	Personal Processes and Medicine to continually improve longevity	Now-1000 years
Intelligence	Human Intelligence	Genetic Engineering to dramatically improve human intelligence	25 Years to 200 Years
Manufacturing	3D Printing	3D Printing in all phases of manufacturing	Now-300 Years
Military	Nuclear Proliferation	Small Unstable Powers and small Nuke Wars—hypersonic aerospace vehicles	Now-150 Years
Nano Technology	Nano Robots	Nanotech used for medicine, war, construction, and more	Now-500 Years
Robotics	Robotics Of All Types	Replacing repetitive tasks and construction, including uses at home	Now-250 Years
Construction	Nest Cities	Horizontal Elevators and all types of movements in cities	Now-300 Years
Space	Space Travel & Colonies	Exploration and Settlement	Now-10,000+ Years
Transportation	Flying Cars	Cars using electric rotors and computer control for personal transportation and Hypersonic airplanes.	5 Years to 100 Years
Transportation	Reusable Rockets	Happening now and continued reductions in cost should continue	Now-200 years
Transportation	Space Elevator	Building a Usable Space Elevator	200-1000 years

Here are some examples of previous paradigm shifts:

Crazy, Strange, and Disruptive Ideas

4.1 Tectonic Plates:

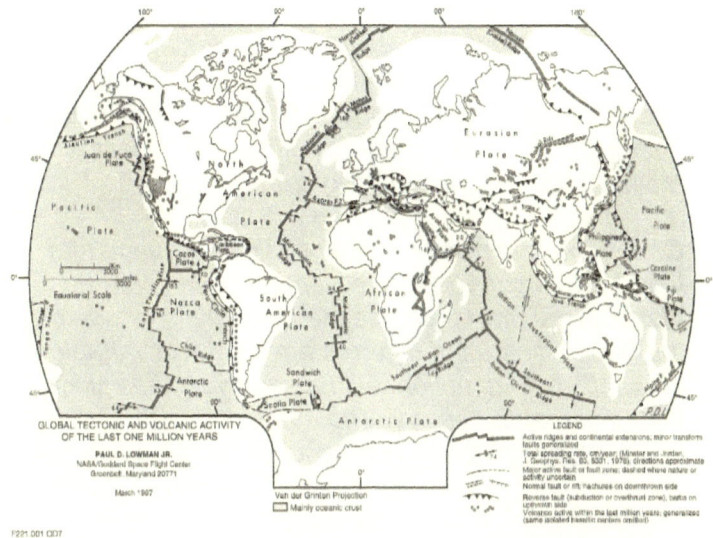

When I was growing up we knew there were lots of Earthquakes all over the world but didn't have any idea about there being a pattern. Also, when you looked at a world map the continents seemed to fit together but it was considered coincidence. An example would be how South America and Africa fit together if the Atlantic wasn't separating them.

Then came the idea of continental drift and plate tectonics. Here is an article on the history of the development of this theory:

> *Despite much opposition, the view of continental drift gained support and a lively debate started between "drifters" or "mobilists" (proponents of the theory) and "fixists" (opponents). During the 1920s, 1930s and*

Crazy, Strange, and Disruptive Ideas

1940s, the former reached important milestones proposing that convection currents might have driven the plate movements, and that spreading may have occurred below the sea within the oceanic crust. Concepts close to the elements now incorporated in plate tectonics were proposed by geophysicists and geologists (both fixists and mobilists) like Vening-Meinesz, Holmes, and Umbgrove.

One of the first pieces of geophysical evidence that was used to support the movement of lithospheric plates came from paleomagnetism. This is based on the fact that rocks of different ages show a variable magnetic field direction, evidenced by studies since the mid–nineteenth century. The magnetic north and south poles reverse through time, and, especially important in paleotectonic studies, the relative position of the magnetic north pole varies through time. Initially, during the first half of the twentieth century, the latter phenomenon was explained by introducing what was called "polar wander" (see apparent polar wander), i.e., it was assumed that the North Pole location had been shifting through time. An alternative explanation, though, was that the continents had moved (shifted and rotated) relative to the North Pole, and each continent, in fact, shows its own "polar wander path". During the late 1950s it was successfully shown on two occasions that these data could show the validity of continental drift: by Keith Runcorn in a paper in 1956, and by Warren Carey in a symposium held in March 1956.

The second piece of evidence in support of continental drift came during the late 1950s and early 60s from data on the bathymetry of the deep ocean floors and the nature of the oceanic crust such as magnetic

Crazy, Strange, and Disruptive Ideas

properties and, more generally, with the development of marine geology which gave evidence for the association of seafloor spreading along the mid-oceanic ridges and magnetic field reversals, published between 1959 and 1963 by Heezen, Dietz, Hess, Mason, Vine & Matthews, and Morley.

Now that plate tectonics is well accepted, we understand why there are volcanos circling the Pacific Ocean. These volcanos appear at the boundaries of the plates.

We also understand that mid ocean ridges often represent an upwelling of the continents and this has led to the discoveries of hotspots and deep sea life based on these hotspots.

This disruptive concept and paradigm change has led to a much better understanding of our world and has been the basis for even newer discoveries.

Crazy, Strange, and Disruptive Ideas

Crazy, Strange, and Disruptive Ideas

4.2 The Internet & Smartphones

The history of the Internet begins with the development of electronic computers in the 1950s. Initial concepts of wide area networking originated in several computer science laboratories in the United States, United Kingdom, and France.

The US Department of Defense awarded contracts as early as the 1960s, including for the development of the ARPANET project, directed by Robert Taylor and managed by Lawrence Roberts. The first message was sent over the ARPANET in 1969 from computer science Professor Leonard Kleinrock's laboratory at University of California, Los Angeles (UCLA) to the second network node at Stanford Research Institute (SRI).

Packet switching networks such as the NPL network, ARPANET, Tymnet, Merit Network, CYCLADES, and Telenet, were developed in the late 1960s and early 1970s using a variety of communications protocols. Donald Davies first demonstrated packet switching in 1967 at the National Physics Laboratory (NPL) in the UK, which became a testbed for UK research for almost two decades. The ARPANET project led to the development of protocols for internetworking, in which multiple separate networks could be joined into a network of networks.

The Internet protocol suite (TCP/IP) was developed by Robert E. Kahn and Vint Cerf in the 1970s and became the standard networking protocol on the ARPANET, incorporating concepts from the French CYCLADES project directed by Louis Pouzin. In the early 1980s the NSF funded the establishment for national supercomputing centers at several universities, and provided interconnectivity in 1986 with the NSFNET project, which also created network access to the supercomputer sites in the United States from research and education organizations. Commercial Internet service providers (ISPs) began to emerge in the very late 1980s. The ARPANET was decommissioned in 1990. Limited private connections to parts of the Internet by officially commercial entities emerged in several American cities by late 1989 and 1990, and the NSFNET was decommissioned in 1995, removing the last restrictions on the use of the Internet to carry commercial traffic.

In the 1980s, research at CERN in Switzerland by British computer scientist Tim Berners-Lee resulted in the World Wide Web, linking hypertext documents into an information system, accessible from any node on the network. Since the mid-1990s, the Internet has had a revolutionary impact on culture, commerce, and technology, including the rise of near-instant communication by electronic mail, instant messaging, voice over Internet Protocol (VoIP) telephone calls, two-way interactive video calls, and the World Wide

Crazy, Strange, and Disruptive Ideas

Web with its discussion forums, blogs, social networking, and online shopping sites.

The result of the incredible internet and World Wide Web is still shaking our world. And if you add Smartphones into the mix, we have an extended paradigm shift of Smartphones able to do things we never did before with the Internet and new technologies like GPS.

Think about how the internet, World Wide Web, smartphones, and GPS have all come together to create an incredible worldwide paradigm shift in the way people live their lives.

Here are just a few of the dramatic changes caused to civilization by the integration of these technologies:

1) Retail stores are closing as more and more people do most of their shopping online.
2) Everyone has cell phones which can surf the web and use many web services wherever they are.
3) Many, many services are offered entirely over the web instead of in person. This even includes medical diagnosis by Doctors in some remote places.
4) Movies are streamed over the web to individuals and homes which has reduced audiences seeing movies at the theater and the video rental business died.
5) The music industry was destroyed and now most people get their music from individual music downloads on the web.
6) Fisherman in poor African villages use their phones and web connections to find where to take their fish for the best market prices.
7) People use their cell phone mapping to find how to get everywhere. Something impossible when I grew up

Crazy, Strange, and Disruptive Ideas

when we used to use paper maps. (Who under 25 can even read a paper map?)
8) Order transportation services using your Smartphone like Uber and Lyft.

What incredible changes the Internet, Smartphones, and related technologies have initiated in our lives. I grew up before the internet and cell phones existed so I know how dramatic the changes have been.

I remember in about 2011 when I was walking at night with my son in Philadelphia looking for a restaurant. We were using my IPhone's mapping capabilities. I saw lots of other people doing the same thing. This is when I became aware that the world had changed dramatically.

Crazy, Strange, and Disruptive Ideas

4.3 Oil and Gas Fracking

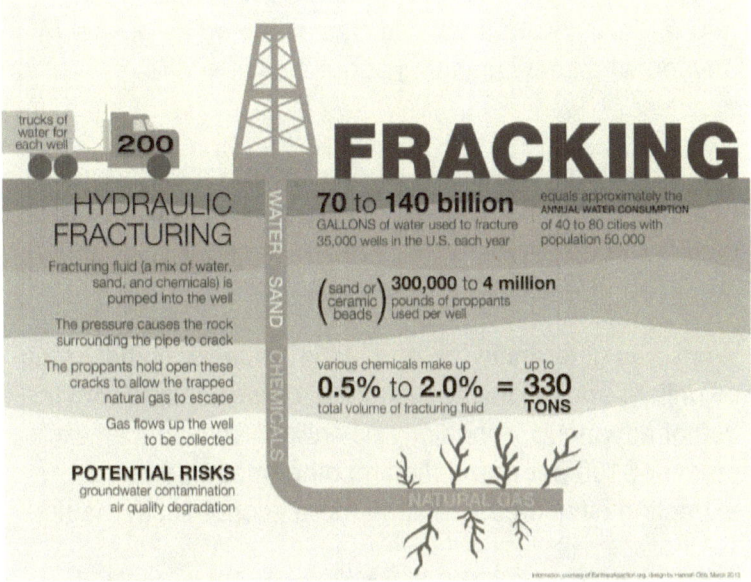

I can remember growing up in the nineteen sixties and seventies which had several big events related to oil and gas shortages and the energy crisis. In the nineteen seventies two major events occurred:

First, was the oil embargo from the OPEC nations because they were upset about the 1973 Israeli and Arab War. I remember sitting in gas lines for hours in the middle of the winter to get a few gallons of gasoline for my parent's car. This taught me how fragile our system based on Oil and Gas was.

In 1973, partly in reaction to the Arab oil embargo, the United States built the Alaskan oil pipeline to the north

Crazy, Strange, and Disruptive Ideas

slope of Alaska. At that time we considered that the United States was at "Peak Oil" and that oil would only become much more expensive and of limited availability in the future. This view of the world was accepted as gospel by anyone who knew anything about the Oil and Gas business.

The concept of fracking has been known for over one hundred years and was sometimes applied to water wells by blowing up explosives in them to crack the rock and produce more water.

What was different in the nineteen nineties was that some oil drillers applied this technique to oil wells along with the use of advanced technology oil drilling. As oil prices went up over $100 per barrel, finding other methods of extracting oil and gas became more economically viable.

The success of oil and gas fracking been a worldwide disruptive technology which has made the United States the largest producer of oil and gas in the world, lowered the worldwide price of oil, and totally disrupted what everyone thought was a basic truth about us having limited resources to get oil and gas from the Earth.

Now, we are much less dependent on the countries in the Middle East so it has helped our stability and the number of alternatives available if the flow of oil from that area is disrupted.

Crazy, Strange, and Disruptive Ideas

5.0 Potential Paradigm Disruptions

There are many potentially disruptive changes which could alter the world. This subject is discussed in a lot more detail in my book "Future Predictions by an Engineering and Seer". I have also listed many potential paradigm changes there along with many predictions for the future.

Remember that disruptive changes are by definition hard to predict because they don't follow a pattern and occur from unexpected sources with results nobody would ever predict.

Here are some summary tables of potential paradigm shifts discussed in my other book on the Future:

Far Out Possible Paradigm Shifts:

Focus	Paradigm Shift Type	Description	Prob	Time Frame
Gravity	Anti-Gravity	Use of Anti-gravity technology for transportation and in phases of society	80%	50 Years to 200 Years
Paranormal	Mass Psychic Abilities	That Psychic abilities will be generally accepted and people trained in their usage	95%	Now to 10,000 Years
Zero Point Energy	Zero Point	Mass Zero Point Energy production from the vacuum	10%	500 Years

Crazy, Strange, and Disruptive Ideas

Technology Driven Probable Paradigm Shifts:

Segment	Paradigm Shift Type	Description	Timeframe
Biology	CRISPR Technology	Large Scale Disease and Biology changes	Now-200 Years
Digital Currency	Crypto Currencies	Digital Currencies independent of any state	Now to 100 Years
Education	Remote Learning	A gradual shift towards most learning online	Now to 100 Years
Farming	Hydroponics	Large scale increased food production	Now-100 Years
Power	Fusion Power	Fusion Power and then miniaturization	30 years-300 years
Health & Medicine	Increasing Longevity	Personal Processes and Medicine to continually improve longevity	Now-1000 years
Intelligence	Human Intelligence	Genetic Engineering to dramatically improve human intelligence	25 Years to 200 Years
Manufacturing	3D Printing	3D Printing in all phases of manufacturing	Now-300 Years
Military	Nuclear Proliferation	Small Unstable Powers and small Nuke Wars—hypersonic aerospace vehicles	Now-150 Years
Nano Technology	Nano Robots	Nanotech used for medicine, war, construction, and more	Now-500 Years
Robotics	Robotics Of All Types	Replacing repetitive tasks and construction, including uses at home	Now-250 Years
Construction	Nest Cities	Horizontal Elevators and all types of movements in cities	Now-300 Years
Space	Space Travel & Colonies	Exploration and Settlement	Now-10,000+ Years
Transportation	Flying Cars	Cars using electric rotors and computer control for personal transportation and Hypersonic airplanes.	5 Years to 100 Years
Transportation	Reusable Rockets	Happening now and continued reductions in cost should continue	Now-200 years
Transportation	Space Elevator	Building a Usable Space Elevator	200-1000 years

In the next few chapters we will review just a few of the possible paradigm shifts which are possible.

Crazy, Strange, and Disruptive Ideas

5.1 Making Everyone Tiny

Lots of experts worry about world overpopulation and overuse of natural resources. There is a solution—**Just make everyone tiny!**

There is a historical example from the Homo Floresiensis species in Indonesia. Those people lived up to 50,000 years ago and were only three feet and seven inches tall.

We could use our knowledge of genetics to breed smaller and smaller people. Who knows maybe we could end up with people who are only six inches tall.

Think of how much we would save on resources! Cars which are only two feet long, food for one current family might feed thirty families.

It would also cost a lot less to send people into space. The future is bright for a tiny humanity!

Crazy, Strange, and Disruptive Ideas

Crazy, Strange, and Disruptive Ideas

5.2 Cryptocurrency Revolutions

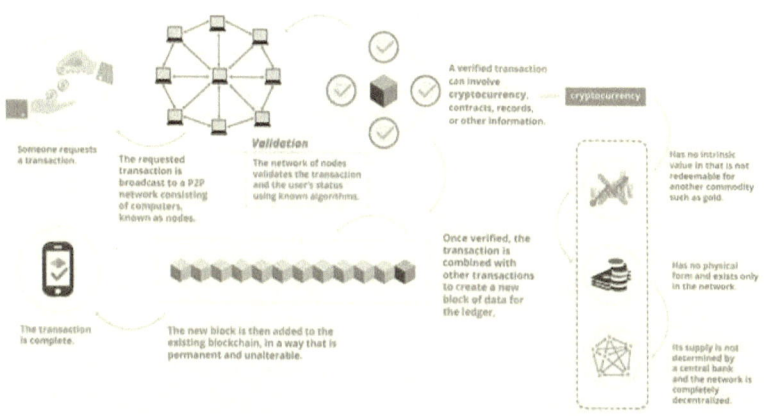

Money in the twentieth century evolved so that each country had their own currencies and there were a few reserve currencies in the world which other countries based the value of their currencies on. The United States and the Eurozone are the two major reserve currencies in the world today.

Cryptocurrencies are a major revolution in the use and formation of money since 2009. Cryptocurrencies are not dependent on the value of the currency in any one country.

It's too early to decide if cryptocurrencies will survive in the long run, but here are some reasons why people might want to use them instead of regular currencies:

• ***Fraud-proof:*** *When cryptocurrency is created, all confirmed transactions are stored in a public ledger. All identities of coin owners are encrypted to ensure the legitimacy of record keeping. Because the currency is*

Crazy, Strange, and Disruptive Ideas

decentralized, you own it. Neither government nor bank has any control over it.

Identity Theft: *The ledger ensures that all transactions between "digital wallets" can calculate an accurate balance. All transactions are checked to make sure that the coins used are owned by the current spender. This public ledger is also referred to as a "transaction blockchain". Blockchain technology ensures secure digital transactions through encryption and "smart contracts" that make the entity virtually unhackable and void of fraud. With security like this, blockchain.*

Instant Settlement: *Blockchain is the reason why cryptocurrency has any value. Ease of use is the reason why cryptocurrency is in high demand. All you need is a smart device, an internet connection and instantly you become your own bank making payments and money transfers.*

Accessible: *There are over two billion people with access to the Internet who don't have rights to use to traditional exchange systems. These individuals are clued-in for the cryptocurrency market*

You are the owner: *There is no other electronic cash system in which your account is owned by you.*

In the future currencies supported by individual countries might go away and fully independent cryptocurrencies might rule the finances of the world.

Crazy, Strange, and Disruptive Ideas

5.3 Direct Voting

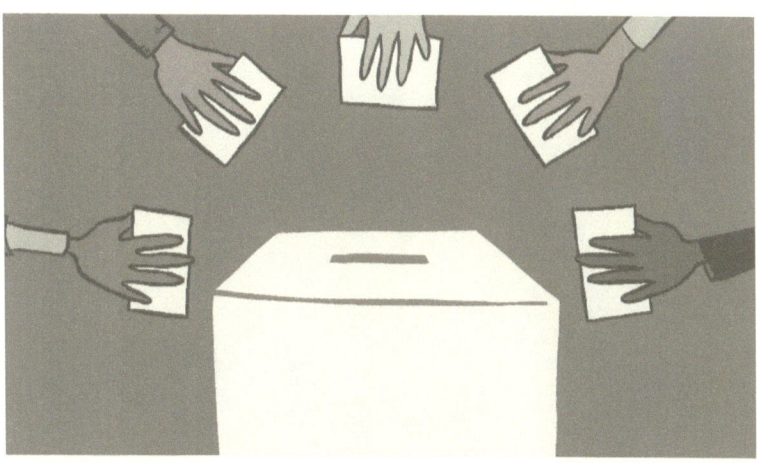

The technology for direct voting by citizens over the internet has been available for twenty years or more.

The problems of making it available to people are political not technological.

Everyone could have a special voter ID produced from government records of social security or other records which are very likely to be citizens. Almost every American now has access to the internet through computers or their phones.

This process could be made much more error and fraud free than the current methods of using voting machines to record votes or paper ballots.

The problem is that some politicians have their own reasons for not wanting to have accurate voter counts. Ove time I believe that governments will become more trusting

Crazy, Strange, and Disruptive Ideas

of online access processes for simpler and more direct voting of citizens.

Crazy, Strange, and Disruptive Ideas

5.4 All Universities Online

How heretical of me to make this statement. I did get a four year degree from the oldest engineering school in the country. When I went to school in the nineteen seventies only about twenty percent of the United States population had four year college degrees.

However, now the percentage of the U.S. population with a four year degree is over 33% and most people agree that our youth needs one to get a good job.

But the job market has changed in the U.S. since I was going to school. Back then a B.S. or B.A. degree almost guaranteed a job. This is no longer the case. Now new graduates must take intern jobs to have a good chance at a full time job after they graduate.

Also, the unbelievable debt which students incur after four years in college is something which people my age would never have worried about. Schools didn't have the option of keeping their prices up for tuition and expenses at the expense of the students going there. Student loans are

Crazy, Strange, and Disruptive Ideas

one of the biggest scams in history. Many former students will never be able to pay off their student loans which continue to grow with interest.

Also, the development of courses and degrees available online means that an inexpensive replacement for physical campuses is in the process of being development right now. As physical campus costs rise and online teaching becomes even better with advanced information technology, then the need for physical university locations for students will have to decrease.

The value of certain degrees also becomes less and less as certifications for different professions become more and more important. Today it is possible to earn an income of over $100,000 per year just by getting the necessary certifications, without ever have a college diploma to a person's name.

So the natural trend in education seems to be to let physical college and university campuses close down and used improved remote educational options for most students.

In fact, you could argue that a college degree is obsolete for many persons. Take the persons who started some of today's biggest tech firms. Bill Gates of Microsoft, Inc., and Mark Zuckerberg of Facebook were both university drop outs because they saw options to build businesses and careers which could not be done by pursuing a university education and working for somebody else.

Crazy, Strange, and Disruptive Ideas

My experience with colleges and universities has to do with the benefits of advanced degrees. I got my B.S. in Engineering Science from Rensselear Polytechnic Institute which is a top school, because I wanted to learn about science and engineering's understanding of our world. I certainly did gain that knowledge and it was useful to me.

Later on I decided to work on an M.B.A. while I was working for General Electric in Des Moines, Iowa. Having completed about the half the degree, I never finished it because I was transferred to a different state by General Electric. Although I took additional course towards an M.S. in Computer Science at University of Houston, and UCLA, I never completed that degree because as I became a business owner and entrepreneur it became less and less useful.

The other main interest if mine was Spirituality and the Paranormal. These are things mainly learned by experience and I never found a degree program which would teach about these thing experientially.

Crazy, Strange, and Disruptive Ideas

Crazy, Strange, and Disruptive Ideas

6.0 Summary

Have I given you some new ideas to think about and ponder?

I've tried to present the reader with some strange or startling ideas which might be nuts, or might offer entirely new paradigms for the future.

If you want to learn more about potential disruptive changes you might want to read my book "Future Predictions By an Engineer and Seer" to get a comprehensive treatment of the subject of new paradigms and potential changes in different industries and government.

Has this book opened your mind at all to what might be possible? Do you think I'm nuts or off my rocker? Your opinion will not bother me if you do as long as I don't get locked up. LOL

I hope you enjoy your future and can use your now more open mind in productive ways to enhance your life.

Marty Ettington

2019

Crazy, Strange, and Disruptive Ideas

Crazy, Strange, and Disruptive Ideas

Bibliography

1. Agartha. *https://www.paleoaliens.com/event/agartha/Agartha_Indians.html*. [Online]

2. Stories of a Hollow Earth. *https://publicdomainreview.org/2011/10/10/stories-of-a-hollow-earth/*. [Online]

3. https://futurism.com/singularity-explain-it-to-me-like-im-5-years-old. *futurism.com*. [Online]

4. https://www.space.com/39510-are-we-living-in-a-hologram.html. *space.com*. [Online]

5. https://www.brandeis.edu/now/2018/march/hologram-qubits-headrick.html. *www.brandeis.edu*. [Online]

6. https://en.wikipedia.org/wiki/Plate_tectonics#Development_of_the_theory. *wikipedia.org*. [Online]

7. https://economictimes.indiatimes.com/industry/banking/finance/5-reasons-why-you-should-go-for-cryptocurrency/articleshow/61184608.cms. *https://economictimes.indiatimes.com*. [Online]

8. https://www.watkinsmagazine.com. *https://www.watkinsmagazine.com/meeting-our-galactic-neighbours-2*. [Online]

9. https://www.universetoday.com/140847/theres-a-surprising-amount-of-life-deep-inside-the-earth-hundreds-of-times-more-mass-than-all-of-humanity/. *https://www.universetoday.com.* [Online]

10. https://www.livescience.com. *https://www.livescience.com/64272-carbon-mass-in-earth-deep-biosphere.html.* [Online]

11. http://personal-longevity.com. *Personal Longevity.* [Online]

www.ingramcontent.com/pod-product-compliance
Lightning Source LLC
Chambersburg PA
CBHW021834170526
45157CB00007B/2802